Planet Dexter's

Crystal Factory

The World's Craziest Crystal Kit Ever

Irene Trimble

AF677766

PLANET DEXTER®

A Satellite of Addison Wesley Longman

Troll

Many of the designations used by manufacturers and sellers to distinguish their products are claimed as trademarks. Where those designations appear in this book and Addison Wesley Longman was aware of a trademark claim at the time of publication, the designations have been printed in initial capital letters.

Set in various sizes and weights of Rotis Semi Sans, Triplex Serif, Madrone and Black Oak.

Books from Planet Dexter are available for bulk purchases by corporations, institutions, and other organizations. For more information, please contact: The Corporate, Government, and Special Sales Department, Addison Wesley Longman, Reading, MA 01867, 1-800-238-9682.

The author wishes to thank Dr. David Dows, Dr. Henry Kaplan, Jeanne Honecutt, Rene Taylor, Danielle Franks, Don Kent Trimble, and Liz Doyle.

Book design: Skolos/Wedell, Inc.

Illustration: Jack Keely

HEY! Get a Load of This!

No candy, no television,

no video games,

no Sega or Nintendo.

No music, no rides to school, no sugar on your cereal or salt on your fries. **Holy guacamole!** Can you imagine a world that's this dull? Who'd want to live there?

Luckily, we don't have to live in a world like that because of crystals.

Crystals rule. They are a totally necessary part of all of the things listed above (and lots of other things we DIDN'T list above).
And **SHAZAM!**
With this book you're going to make some!

You probably should know, before we start, that these are some mellow recipes.

The crystal stuff you can make with this book is very laid back.

You don't have to get the measurements exactly right, and you don't have to keep the crystal stuff totally stock still. It's OK if you accidentally bump into it or move it or poke it. (Try to avoid doing all this if you can, though. These babies are fragile.)

Another thing: sometimes there will be a **warning,** or you'll need someone of the **adult persuasion to help you out.**

Don't blow off these warnings –
after all, scientists always have partners.

It's a Fact!!
Burping while lying on your back is a lot harder than burping while you're sitting or standing.

Anyway,

enjoy the book. Go grow some crystals! And hey, if you have any questions (for the author or for the editors at Planet Dexter), you can write to us. Really! Here's the address:

Planet Dexter
One Jacob Way
Reading, MA 01867

You can also fax us at: (617) 944-8243. Or, send us an e-mail message via the Internet at pdexter@aw.com or on America Online, at PDexter. OK? We'd really like to hear your questions or comments, and we'd especially like to hear any compliments or complaints that you might have.

Thanks.

Finally, one thing to know about these experiments is that some of them take a few days. So you'll have time to kick back, relax, and plot your next move.

Keeping Track of Your Experiments

You may not know this, but all scientists keep journals. The reason they do is that if they discover something great, they can go back and figure out how they got there. Most scientists set up their journals in the same general format:

1 **The name of the experiment**

2 **The date and time (Growing Time) —**
when it was started and how long it took

3 **Goal or Hypothesis (Mission) —**
what the scientist hoped to see or accomplish

4 **Materials (The Secret Formula) —**
what ingredients were used and how much

5 **Procedure (Mixing the Magic Potion) —**
what the scientist did

6 **Observations (Spying on Your Crystals) —**
what the scientist saw

The great thing about Planet Dexter's Crystal Factory is that it turns into a science journal if you want it to. Most of the experiments have a section where you can record your observations and the conditions (like the weather or which room you did it in) under which you created the experiment. Then the instructions turn into the first portion of your journal, reminding you of what the experiment was and what you put into it.

It's a Fact!
An automatic spaghetti-spinning fork was patented in 1950.

Here are some questions you might want to answer in the journal section "Spying on Your Crystals."

1. When and where do the crystals first appear?
2. What is the shape of the crystal?
3. How do the crystals you've grown compare to the ones you dissolved in your solution?
4. Do these crystals refract light so that they make rainbows?
5. How could other factors have affected the growth of your crystals (consider things like the weather and the family dog)?
6. What can you make out of your crystals? How about jewelry, a prism, or bath salts?

The observation section of your journal can also be a place for drawing or sketching your results or making predictions. Or, you can ignore the science aspect and use the journal space to list your best friends (in order of preference), or all of the roller coasters you've ever been on, or what presents you want for your birthday. **Feel free to write whatever you want.**

DID YOU KNOW?

That a dentist uses a diamond (diamonds are crystals) drill on the enamel of your tooth to prepare it for a filling? The diamond drill cuts through enamel faster than any other drill can. It's a good thing, too. Otherwise you might spend days at the dentist! Nightmare!

BEFORE YOU BEGIN THE EXPERIMENTS, DON'T FORGET TO FOLLOW THESE STEPS:

1. Read the experiment through.
2. Get out all the materials you'll need (The Secret Formula).
3. Ask an adult to help you if you need to use the stove.
4. Try to use the container size that the directions recommend (or as close a size as you can find), so that your crystals turn out like they should.
5. Do not eat or taste any crystals or crystal growing solution, except for Rock Candy.
6. Clean up your work area after your experiment is done.
7. Always wash your hands when you are finished.

Here's the secret language of crystals. You'll need these words to get through the world of crystals. Good luck, and remember, forewarned is forearmed.

ATOM

The smallest particle of an element that can take part in a chemical change without being permanently changed itself

CRYSTAL

A solid that has flat sides and a symmetrical form due to the arrangement of its atoms and molecules

MOLECULE

The smallest particle of a compound that retains all the chemical properties of that compound (H_2O is a molecule)

SOLUTE

A substance dissolved in something else

STALACTITE

Stone formation that resembles an icicle hanging from the roof of a cave, formed by water seeping through and dissolving small amounts of limestone or similar rock

STALAGMITE

Cone-shaped stone formation on the floor of a cave, formed by the dissolved minerals in water dripping from a stalactite

SUBLIME

To change a solid substance into a gas or a gas into a solid

CRYSTALLOGRAPHY

The branch of physics that studies the geometric forms, internal structures, and other properties of crystals

DISSOLVE

To pass into solution, to be absorbed, usually into a liquid

INTRINSIC PROPERTY

Basic attribute or feature of something

EVAPORATE

To change from a liquid to a gas

SOLUTION

A mixing of two or more substances so that the molecules of one are evenly dispersed through the others

SOLVENT

The dissolving substance of a solution, usually a liquid if the solution consists of a solid and a liquid

SUBSTANCE

Material of which something is made

SUPERSATURATE

To cause a solvent to hold more of a solute than it normally could, usually by warming the solvent

SYMMETRY

Balanced proportion either by a balance of halves on opposite sides of a line or by the balance of parts around a central point or axis

If you want to know how crystals form, read this. Otherwise, go on ahead and jump into those experiments willy-nilly. That's cool: you can always come back to this page later.

Now, for all you friendly and inquisitive types who stuck around . . .

The easiest way to explain how crystals grow is to use the Rock Candy experiment as an example.

(You can check it out. It's on page 18.)

OK. Here goes: when a solute (sugar) is dissolved in a solvent (water), it creates a solution (sugar water). When no more solute (sugar) can be dissolved in the solution, you have what is called a saturated solution. **Phew!** Try saying that ten times fast!

Now, to get crystals to grow, you have to get even MORE sugar to dissolve in the saturated solution. You CAN DO THIS by heating the saturated solution.

HOW CRYSTALS GROW

When more sugar has been dissolved in the hot solution and you begin to cool it back down, you have what is called a supersaturated solution. ("Super" means "above," so here it's saturated with "**above** the usual amount" of sugar it can hold.)

A Helpful Hint that Parents DO NOT Want You To Know

You can easily wreck a slice of healthy, "really-good-for-you," dark-and-seedy bread and revive several stale and "really-bad-for-you," hard marshmallows by putting them together in a plastic bag for three days; after which you'll have stale bread and soft marshmallows. Great!!!

As the hot, supersaturated solution cools, **CRYSTALS** begin to form. The sugar molecules that were packed into the hot solution reform into solid crystals because there's just no room for them to remain dissolved in the cooled, supersaturated solution.

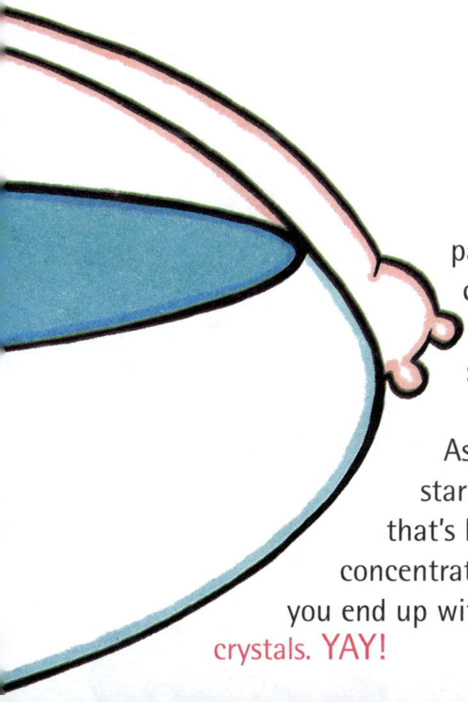

As the water from the supersaturated solution starts to evaporate (disappear into the air), the sugar that's left behind becomes more and more concentrated, and crystals continue to grow. After 7 days, you end up with a whole plate of ROCK CANDY sugar crystals. YAY!

What the Heck is a Crystal?

A crystal is a solid with **flat sides,** with all its molecules (little pieces) organized in an **orderly** and repetitive pattern. A crystal is usually made up of only **one kind of molecule.** Most solids are made up of lots of different kinds of molecules.

When a solid is made up of only one kind of molecule, scientists call it a pure substance, or a sample. Salt, for example, is a pure substance made up of sodium chloride molecules and nothing else. That sodium chloride must get pretty sick of its own kind of molecule. **Dullsville.**

And Why do they look like THAT?

Because the crystal is pure (made up of only one molecule throughout), all the molecules within the crystal take the same arrangement with their molecule neighbors (**"Hey, how ya doin'? I'm Ralph from next door. Mind if I sit here?"**). This arrangement is controlled by the shape of the molecule and by its interaction with its neighbors.

Here's an example. Imagine that you have a lot of blocks (cubes) that represent the molecules of a crystal. If you begin stacking the blocks in layers to create your imaginary crystal, eventually the identical layers of small cubes will be stacked high enough to create a larger cube. Cool, huh? Think about this: if it takes eight cubes to make up the next bigger cube, then how many cubes do you need to make the next largest cube after that?

MONOCLINIC

There are seven standard crystal shapes, called systems (check out these diagrams). These systems are (drumroll):

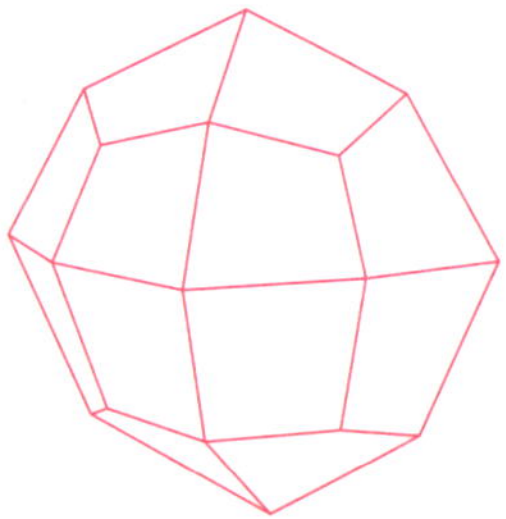

CUBIC

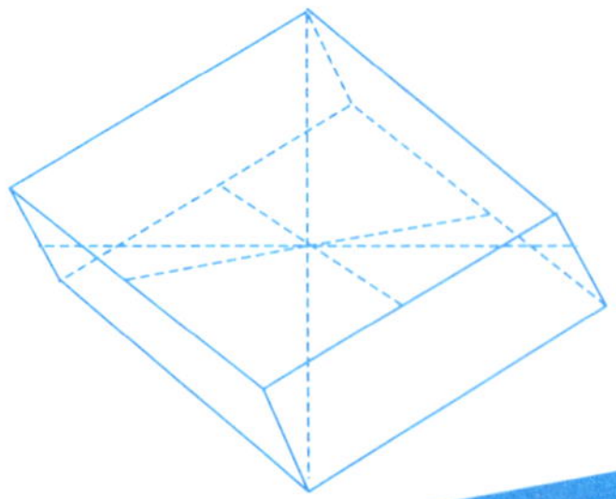

RHOMBOHEDRAL

DID YOU KNOW

That the first synthetic ruby (rubies are crystals) was produced in 1937 by fusing alum and chromium oxide at a high temperature? Today, synthetic rubies are made into tubes for creating lasers.

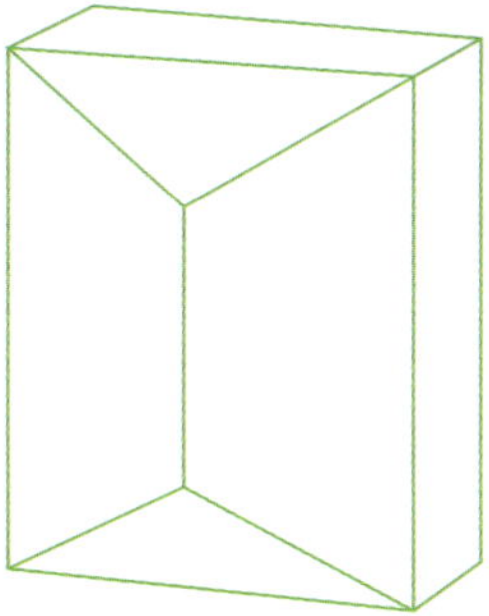

ORTHORHOMBIC

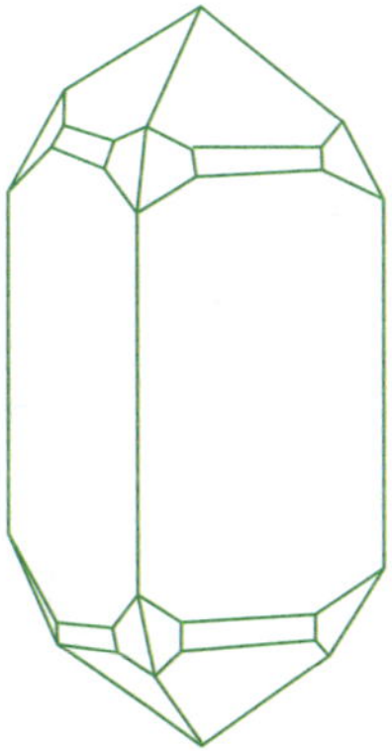

TETRAGONAL

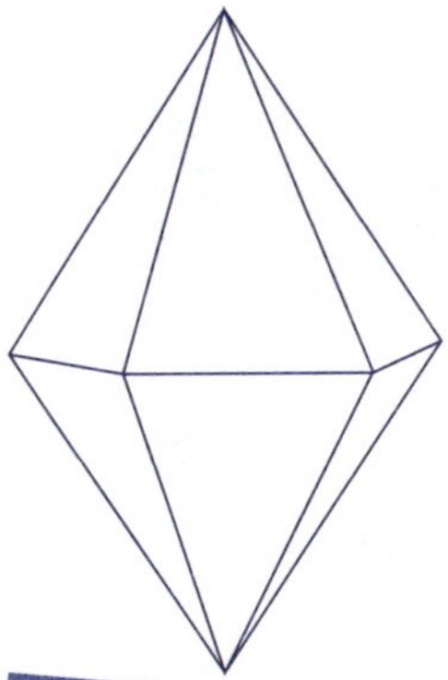

HEXAGONAL

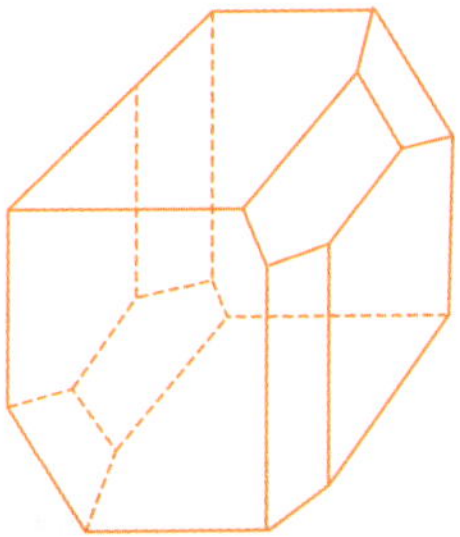

TRICLINIC

Scientists use crystallization information to find things out about substances. Scientists can find out how pure the substance is and, by the shape of the crystal itself, something about the shape of its molecules. Elementary, my dear Dexter!

Rock Candy

MISSION

These crystals are just like rock candy in the store. It's the easiest rock candy you'll ever grow. And, this is the only crystal experiment in this book that you can EAT!!!

THE SECRET FORMULA

One large plastic container
(like a square, 1.3-gallon Rubbermaid™ container)
Do not use a glass bowl for this experiment; it might break, and the crystals will not pop out.

One large (4-quart) saucepan

2 cups of water

4 cups of sugar

Food coloring
(optional)

Wax paper

Spoon

It's the Law!!
In Denver, Colorado, it is illegal for children to do handstands on sidewalks because it might scare horses.

Mixing The Magic Potion

1. **Ask a friendly adult to help you use the stove.**
2. Place **2 cups of water** in a large saucepan.
3. Add **4 cups of sugar** and stir for a minute to get as much sugar as possible to dissolve.
4. Place the saucepan on the stove and turn the heat to **medium high.**
5. **Stir** the solution over heat until all the sugar is dissolved.
6. If you have any, add some **food coloring** until you like the color. Stir again.
7. Take the saucepan off the heat. If you are using a Rubbermaid™ or Tupperware™ container, you can pour the hot liquid from the saucepan directly into it. If you're not sure, let the hot liquid cool in the saucepan; then pour it into your plastic container.

NOTE: Larger crystals grow if the supersaturated solution cools slowly. So if possible, place your container in a warm spot.

Crystals will form on the surface of the solution in a few days. Do not remove these. Instead, sink them to the bottom with a spoon. This procedure allows the evaporation process to continue.

In 7 days or so, the bottom of your plastic container will be covered with a thick layer of sugar crystals (rock candy).

8 **Turn the container over in a sink** and let the excess sugar solution drain completely. This step takes about 2 hours.

9 When the surface of the crystals is dry, push on the sides and bottom of the plastic bowl, and the crystals will break up into **chunks.**

10 Put the crystals on a piece of wax paper and let them dry completely. The more the crystals dry, the more they sparkle! Break them into smaller pieces if necessary.

OTHER ROCK CANDY IDEAS

Just for fun, get inexpensive pin backs at a craft shop. Glue on your rock candy and turn your crystals into **"real gems" you can wear!**

For a fabulous treasure hunt, make four batches of rock candy in four different colors. Place the rock candy in a treasure box along with some chocolate gold coins. Hide the jewel-filled box. Make up some clues, and the hunt is on!

Compare the crystals you've grown to the size and shape of some plain granulated sugar. How many times larger is one of your crystals than one grain of sugar?

JOURNAL OBSERVATIONS

Date Started: ____________

Date Completed: ____________

Conditions: ____________

Colors: ____________

Weird Stuff: ____________

Other Observations: ____________

Strange or Funny Things You Thought of While Doing or Watching the Experiment: ____________

Bath Crystals

MISSION

Here's how to make colorful, perfumed bath salts to give as gifts (adults love this sort of thing) or keep for yourself (Relax! You deserve it!). Your jar of bath salts can be all one color, or you can make layers of colored crystals!

CAUTION: Do not eat the Epsom salts. They're not poison, but they are a fast-acting LAXATIVE! Gross!

GROWING TIME: No time at all

William Howard Taft is the only known United States President to get stuck in the White House bathtub. Really! Soon thereafter, he ordered a new tub, one large enough for four men! (Why would you want four men in one bathtub?)

THE SECRET FORMULA

One small (4-ounce) **glass jar with a lid** or any unique or pretty jar you might want to use (baby food jars work well)

Epsom salts (If you are planning on making lots of bath crystals for gift giving, a 4-pound box of Epsom salts from your pharmacy costs about $2.60 and will fill nineteen 4-ounce baby food jars.)

One sealable sandwich bag for each color of crystal (Ziploc™ works well)

Perfume

Food coloring

Mixing The Magic Potion

1 **Pour Epsom salts** (as much as you need to fill your jar) into your plastic bag. If you are going to make layers of colored crystals, divide the Epsom salts into several bags so that you can dye them different colors.

2 **Add** a few drops of **food coloring**.

3 **Seal** the sandwich bag. **Shake** and knead the bag with your fingers to distribute the color.

4 **Open** the bag and **add one or two drops of perfume** (if you like). Seal and knead the bag again.

5 Empty the bag of crystals onto a piece of wax paper and let them dry for no more than 10 minutes

6 Put the crystals into the jar.
You can make layers of different colors (rainbows!) or put them in all at once (bath confetti!). Put on the lid.

Decoration

Do as much as you'd like to decorate the jars. You can make your own labels, stick some stickers on the jar, paint it, put a piece of tissue or fabric over the top of the jar and tie a colorful piece of yarn or ribbon around the neck.

SPYING On Your Crystals

It's amazing how evenly the food coloring is distributed throughout the crystals. This distribution occurs because crystals made of Epsom salts easily absorb any liquid. In fact, there's a certain amount of water in these crystals to begin with. This crystal's chemical formula is Mg (magnesium) SO_4 (sulfate) $7H_2O$ (water). If crystals made of Epsom salts are left out to dry (exposed to air) for too long, the water, H_2O, will evaporate from them, and they'll turn from crystals into powder.

Macaulay Who?

Jackie Coogan, a child actor, became a millionaire at the age of six. Shirley Temple was a millionaire by age ten.

JOURNAL OBSERVATIONS

Date Started: ____________

Date Completed: ____________

Conditions: ______________________________

____

____

Colors: ______________________________

____

Weird Stuff: ______________________________

____

____

Other Observations: ______________________________

____

____

Strange or Funny Things You Thought of While Doing or Watching the Experiment: ______________________________

____

____

____

____

Simmering Rainbows

MISSION

Clear crystals will appear quickly on the inside of a sealed jar and will undergo many changes over the next few days. When you place the jar in a sunny window, the crystals refract the sunlight into tiny rainbows!

THE SECRET FORMULA

One 1-quart Mason jar

One mothball
(naphthalene)

Small saucepan

Hot running water

GROWING TIME 10 to 15 minutes

CAUTION

Mothballs are not to be eaten or tasted! Do not open the jar during this experiment. Dispose of the jar and its contents without opening when you are done.

Mixing The Magic Potion

1 Put the **mothball** in the **jar** and **screw the lid on tightly.**

2 **Place the jar in a small saucepan.**

3 Place the saucepan in the sink, turn on the water, and let it run until it's as hot as it gets. Then let the water run into the pan to heat the bottom of the jar (don't let the water run over the top of the jar, or the experiment might not work).

4 **Let the jar sit in the hot running water until the mothball has completely disappeared.** This step will take about 10 to 15 minutes.

(See next page for more!)

> In 60 seconds, your blood makes a complete trip through your body.

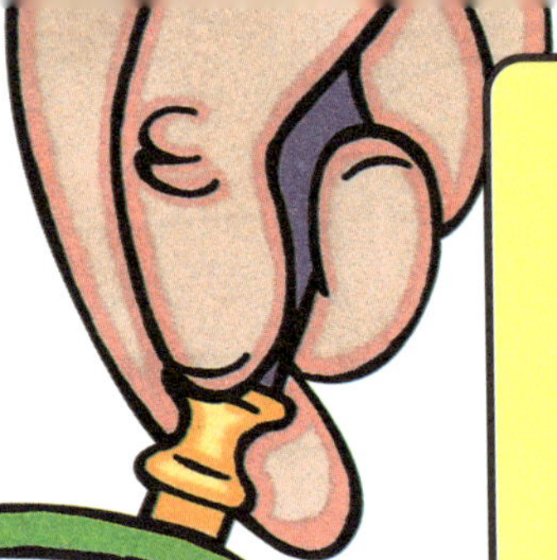

Remove the jar from the saucepan and take a look at the naphthalene crystals inside the jar. **REMEMBER: DO NOT OPEN THE JAR.** Where did the mothball go? It didn't melt. The mothball has sublimed: it changed from a solid directly into a gas without turning into a liquid first. The crystals form when the gas condenses on the cool sides of the jar.

SPYING On Your Crystals

Don't forget to put your jar in a sunny window and watch.

JOURNAL OBSERVATIONS

Date Started: ____________

Date Completed: ____________

Conditions: ____________

____

____

Colors: ____________

____

Weird Stuff: ____________

____

____

Other Observations: ____________

____

____

Strange or Funny Things You Thought of While Doing or Watching the Experiment: ____________

____

____

____

____

Another Example of Subliming a Crystal

CAUTION: Dry ice should be handled with tongs or oven mitts! Dry ice can cause bad burns to your skin. Imagine that! An ice burn!

Maybe you've seen dry ice on MTV. (People like to use it in videos because it looks like smoke or fog, but it's really just cold air. Sorry to ruin the magic for you.) **But did you know that dry ice, which is solid carbon dioxide, is actually a crystal?** Its crystal formation is cubic (see page 16), and its chemical formula is CO_2. Can you guess what causes dry ice to make that spooky, smoky fog? The dry ice crystal, solid CO_2, sublimes into carbon dioxide gas. The fog you see over the dry ice is water vapor condensing as it becomes cold.

If you'd like to try subliming some of your own, dry ice is usually available from your supermarket.

Crystal Fireworks

CAUTION

Be careful with the glass pan. If you accidentally drop it, don't move. Call an adult (who's wearing shoes) to come and save you.

don't eat crystals

GROWING TIME: About 1 hour

MISSION

Here's what will happen: on a flat piece of glass, crystals that look like fireworks (big, explosion-shaped outlines) will begin to form within about 20 minutes. In about 1 hour, the piece of glass will be covered with these crystals.

NOTE:
if you choose to use some food coloring in this experiment, do not use red – it turns muddy brown!

THE SECRET FORMULA

Hot tap water

Two small drinking glasses

One spoon (for stirring)

1/2 teaspoon liquid detergent

Measuring cup

Measuring spoons

4 teaspoons Epsom salts
(magnesium sulfate)

One glass baking pan

Food coloring
(optional)

PREPARING THE PAN

1 In the first drinking glass, put a little less than 1/2 teaspoon of liquid detergent into 1 cup of warm water and stir.

2 Place your glass pan in the sink and pour the soap and water solution over it so it is completely covered. Let it sit for 5 minutes.

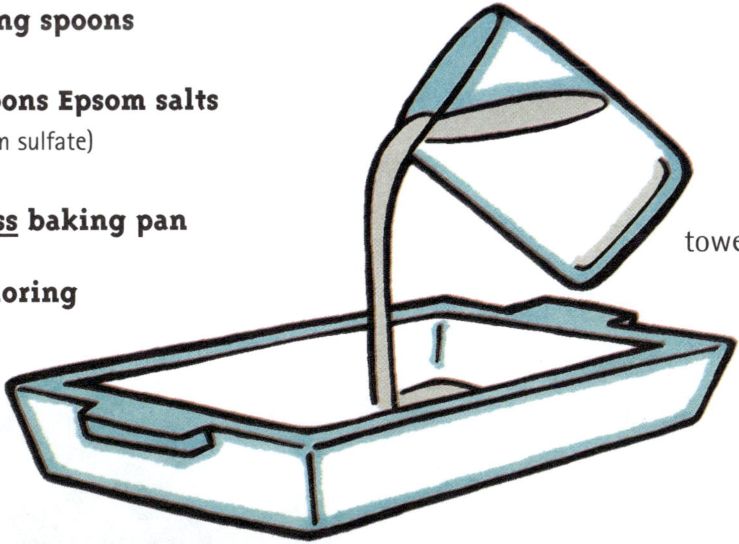

3 Tip the pan to let the solution run out of a corner and into the sink.

4 Place the pan on a paper towel and allow it to dry. The film of soap is what gets your crystal solution to stick to the glass and not bead up.

(go to the next page ⟶)

Mixing The Magic Potion

DID YOU KNOW? that crystals have played a huge role in the development of many medicines? Lots of people take "crystal medicine" every day. Aspirin is a crystal. The reason why aspirin is white is because the little bit of aspirin crystal, acetyl salicylic acid, is pressed with white starch into the form of a pill. (Starch or no starch, they still taste barfy.)

1 **Put 4 teaspoons of Epsom salts in** the other **drinking glass.**

2 **Add 2 tablespoons of very hot tap water.**

3 **Stir to dissolve.** Add a little food coloring if you like and stir again. Pour the solution into the glass baking pan.

4 **Spread the solution to the edges** of the glass by tilting the pan.

5 **Let the solution set for 5 minutes.** Then tip the pan to let some of the solution run into the sink – crystals will grow faster if you do this. In fact, if the crystals aren't growing as fast as you'd like, just let some more of the solution run off.

6 **Sit back and wait.**

In about 15 minutes, begin to check the glass. Where do the crystals appear first? How would you describe the shape of these crystals? Do you think they look like fireworks? What happens if you shine a flashlight through the pan in a dark room? Why do you suppose letting most of the crystal solution run off makes the crystals appear faster?

SPYING On Your Crystals

JOURNAL OBSERVATIONS

Date Started: ______________

Date Completed: ______________

Conditions: ____

____

____

Colors: ____

____

Weird Stuff: ____

____

____

Other Observations: ____

____

____

Strange or Funny Things You Thought of While Doing or Watching the Experiment: ____

____

____

____

____

Cave Dweller Crystals

(This is a tough one)

MISSION

In this experiment, you can create stalagmites and stalactites just like the cave formations described on page 37, but a lot smaller. Only instead of leaky limestone, these stalagmites and stalactites will be formed by a supersaturated solution of Epsom salts dripping from yarn suspended between two jars. The dripping solution evaporates, leaving behind the crystal of Epsom salts. This is a tough experiment; it might be easier to go see the real thing in Kentucky's Mammoth Caves than to get the rate of drip in this experiment just right! But it can be done if you have nerves of steel.

GROWING TIME:
About 4-8 hours, but give it a full day for complete results

THE SECRET FORMULA

1 1/2 cups of Epsom salts

2 cups of water

Saucepan

Two small, same-size, clean, empty **jars**

Two 15-inch pieces of wool yarn

Two paper clips

One piece of wax paper

Spoon for stirring

Mixing The Magic Potion

1 Place a piece of **wax paper on your counter.** Place the **jars** on the wax paper about **4 inches apart.**

2 Put the two pieces of yarn side by side and **twist** them together (see next page). **Attach a paper clip** to each end of the yarn. (You don't have to tie a knot; just clip them to the ends of the yarn.)

3 In a saucepan (ask an adult to help you), bring **2 cups of water to a boil**. **Add 1 1/2 cups of Epsom salts** and stir until it's dissolved.

4 **Thoroughly soak the yarn** in the supersaturated Epsom salts solution.

5 **Hang the string between the two jars.**

(More! ⟶)

6 Pour the supersaturated solution into the jars and fill them to 1/2 inch from the top.

7 Adjust the yarn to hang about 1/2 inch from the top of the jar. The trick is to get the yarn to drip slowly. If the rate of drip is too fast, just raise the yarn a bit so it doesn't hang as low between the jars. If it's too slow, lower it.

What can you do to change the results of this experiment? How do things change when you move the yarn around? What do the results look like?

1,000,000,000 M&Ms are sold on any given day in America; and most of them are brown.

TOTALLY ALL-NATURAL CRYSTALS (EXCELLENT!)

Some of the most incredible-looking crystal formations are stalactites and stalagmites. These formations are made by water that contains a dissolved mineral such as calcite or limestone, which slowly drips from the ceiling of a cave. The water evaporates and leaves behind the crystallized mineral.

These little deposits of minerals eventually build up into big hulking piles that look like giant teeth sticking out of the roof and the floor of a cave (so the cave looks like the mouth of a huge ugly monster who didn't floss). Stalactites hang from the roof of the cave, and stalagmites grow on the floor of the cave (you can remember these definitions because "stalactites hold on tight" and "stalagmites push with all their might").

You can see really cool stalactites and stalagmites in different caves around the world (like Mammoth Caves National Park in Kentucky or Carlsbad Caverns National Park in New Mexico). If you can't make it to see the real thing, you can ask a librarian if the library has books that contain photographs of these formations. It's definitely worth asking. Stalactites and stalagmites are really cool.

JOURNAL OBSERVATIONS

Date Started: ____________

Date Completed: ____________

Conditions: ______________________________

____

____

Colors: ______________________________

____

Weird Stuff: ______________________________

____

____

Other Observations: ______________________________

____

____

Strange or Funny Things You Thought of While Doing or Watching the Experiment: ______________________________

____

____

____

____

Mining Artificial Gems

CAUTION
Do not taste or eat solution of crystals made of Epsom salts.

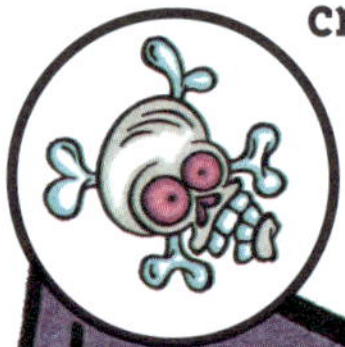

MISSION

These large, clear, jewel-like crystals grow in all sorts of interesting formations at the bottom of a glass bowl.

THE SECRET FORMULA

3 1/3 cups of Epsom Salts

1 tablespoon of sea salt

4 cups of water

Food coloring (optional)

A 2-quart glass bowl

One large saucepan

Measuring spoons

Spoon for stirring

Fork

Wax paper

GROWING TIME:
5 days, longer if you want even larger crystals

Mixing The Magic Potion

1 Ask an adult to help you use the stove.

2 Heat 4 cups of water in a saucepan over high heat **to a full boil.**

3 Add 3 1/3 cups of Epsom salts and **1 tablespoon of sea salt** all at once to the boiling water.

4 Stir the solution over high heat until the Epsom salts and sea salt have dissolved.

5 Remove the saucepan from the heat. Add the **food coloring** if you like, and stir.

6 Pour the solution into the 2-quart glass bowl. Let the bowl stay in a place where it will not be moved while your crystals are growing.

7 In 5 days, or when you've decided your crystals are large enough, take a fork and **gently lift the crystals from the solution.** Place them on the wax paper to dry. It's OK to put your crystals back into the solution to let them continue to grow, if you choose.

The Crystal Ocean

MISSION

GROWING TIME:
7 1/2 to 8 1/2 hours.

In this experiment you can produce two kinds of beautiful crystals in one container. There will be delicate frosty foam at the top of your container, and sometimes, at the bottom, there will be spiky bonus crystals too.

CAUTION
Remember, do not eat or taste the crystals or crystal growing solution

THE SECRET FORMULA

3 3/4 cups of Epsom salts

1 teaspoon of sea salt

2 cups of water

Food coloring (optional)

1.2-quart plastic bowl. (The size of the container is important because the crystal solution must be a certain depth to grow the long crystal spikes.)

One saucepan

Measuring spoons

One spoon for stirring

Fork

Wax paper

Paper towels

Mixing The Magic Potion

1 **Ask an adult to help you use the stove.**

2 Heat **2 cups of water** in a saucepan over high heat to a **full boil.**

3 **Add 3 3/4 cups of Epsom salts** and **1 teaspoon of sea salt** all at once to the boiling water.

4 **Stir** the solution over high heat until the Epsom salts and sea salt are **dissolved.**

5 Let the solution come to a boil once again (you don't have to continue to stir). Then remove it from the heat.

6 **Stir in the food coloring** and **pour** the solution **into** the 1.2-quart plastic **container.**

7 **Allow your crystals to grow** without moving your container for 7 1/2 hours.

(more stuff on the next page)

Wow! Baby blue whales gain ten pounds every hour.

Harvesting The Crystals From Your Mini-Ocean

1 If the solution in your container looks **cloudy** (you can't see through it), then your crystals are **ready to be removed.** If it still looks clear, then give them a little more time to grow.

2 **Place** a piece of **wax paper** and **two paper towels** on the counter next to the sink. Have a fork ready.

3 **If a white crust** has formed on the surface of your solution, use the fork to **remove it carefully** (it's OK if you don't get it all).

4 **Take your container of crystals to the sink** and gently tilt the container to let all the excess solution drain off. Don't turn the container completely upside down, or your crystals will end up in the sink.

5 **Using a fork,** take out a scoop of crystals and **place them on wax paper**. Don't worry if the crystals seem loose. As they dry, new crystals will form to hold them together.

6 **To get the long, spiked crystals** that have grown at the bottom of this container, break the thick bottom layer into large chunks with a fork. Holding these chunks in your hand, place them under running water for about 5 seconds to expose a new layer of crystals (this will make them look shiny). Place them on a paper towel to dry.

SPYING On Your Crystals

It's amazing to see how different these crystals are from the ones in the "Mining" experiment. Not only are these new crystals opaque (that means you can't see through them), but they are also very soft and delicate, unlike the crystals that took longer to grow. Are there any other differences you've noticed?

MAD SCIENTIST

You can make a solution using equal amounts of Epsom salts and water (2 cups of Epsom salts and 2

VARIATION

cups of water plus 1 teaspoon of sea salt) and come up with yet another crystal formation.

JOURNAL OBSERVATIONS

Date Started: ______________

Date Completed: ______________

Conditions: ____

____

____

Colors: ____

____

Weird Stuff: ____

____

____

Other Observations: ____

____

____

Strange or Funny Things You Thought of While Doing or Watching the Experiment: ____

____

____

____

____

Mutant Crystals

MISSION

A powdery salt crystal will begin to form quickly above the solution on the sides of the glass. This creeping crystal process will continue as crystals thicken and grow upward, eventually crawling up to the top of the glass and down the outside, like a mutant crystal slime.

> GROWING TIME:
> First crystals appear within 24 hours. The second, larger crystals appear in about 2 weeks.

While this growth is happening, clear, powerful-looking, jewel-like crystals will form on the bed of salt at the bottom of the glass.

This is a good experiment to demonstrate the intrinsic properties of crystals. Although the two kinds of salt crystals you've grown may look different, close examination (a magnifying glass would be helpful) will show you that they are in fact the same, even though one looks creepy and the other looks powerful and pure.

THE SECRET FORMULA

One clear drinking glass

One spoon (for stirring)

Hot water

Table salt

Food coloring (optional)

Mixing The Magic Potion

1 **Fill the glass one-third full with salt.**

2 **Pour in hot water** until the glass is two-thirds full.

3 **Stir** for a few minutes to get as much salt as you can to dissolve. There will be lots of salt that won't dissolve. That's OK. You want to overload the solution.

4 **Add** your **food coloring** and stir again.

5 **Place the glass on a sunny windowsill** (you get a faster rate of evaporation that way). Choose a spot where you can observe it.

Note: if you don't place the glass in a sunny window, it takes a lot longer to grow the crystals at the bottom.

(more mutant stuff next page)

When you're finished growing the crystals, scrape some of the powdery crystals out of the glass. Then carefully empty out the salt solution and remove the jewel-like crystals from the bottom of the glass.

Compare the two types of crystals for size and shape. Compare the crystals you've grown to some salt from a **salt shaker.** This time you've grown crystals that are larger and smaller than the salt you dissolved in the solution, but both have the same intrinsic properties.

All crystals have certain intrinsic properties, meaning that they maintain or keep their basic characteristics. For example, even if you were to crush a salt crystal, the tiniest particles would still maintain the same taste, the same shape, density, and solubility. That's because all of these particles have the intrinsic properties of salt.

DID YOU KNOW that surgeons use scalpels with blades made of crystals? It's true; in delicate types of surgery, doctors use scalpels with diamond blades because they do not get rusty like some metals.

SPYING On Your Crystals

The undissolved salt will form a bed of salt at the bottom of the glass, where the larger crystals will appear in a few weeks. You can see these crystals by raising the glass to eye level.

JOURNAL OBSERVATIONS

Date Started: ________________

Date Completed: ________________

Conditions: ____

____

____

Colors: ____

____

Weird Stuff: ____

____

____

Other Observations: ____

____

____

Strange or Funny Things You Thought of While Doing or Watching the Experiment: ________________________________

____

____

____

____

POPCORN ROCK

Another cool way to discover nature's crystals is by using a popcorn rock. Popcorn rock was discovered in 1865 by R.D. Barnes in Bountiful, Utah. He found that by placing a popcorn rock in vinegar, you could grow large aragonite crystals that would last for years!

If you want to experiment with popcorn rock, you can try to buy it at science supply stores, science or children's museum gift shops, or other kids' stores (look in the Yellow Pages for places).

If you can't find them anywhere, the science whiz who wrote this book buys hers at:

The Explore More Store
Pacific Science Center
200 Second Avenue North
Seattle, WA 98109
(206)-443-2870

You can write to the people at this store or give them a call. They'll be happy to send you some.

COOKING UP YOUR OWN CRYSTALS

OK, so not everyone likes to go into the medicine chest in the bathroom or the cabinets in the kitchen and root around and dump stuff into bowls and mix it all up. But most people like it. It's fun to make a mess and create goop, and experiment. You can do that with crystal growing too. Check it out: all of this stuff is good for growing crystals.

- Cream of tartar
- Washing soda
- Alum (a spice)
- Epsom salts
- Salt
- Sugar
- Baking soda

The trick is figuring out the **MAXIMUM** (most) amount of this stuff (the solutes) that can be dissolved in a cup of hot water (the solvent). **So, mix it up. Go on, don't be shy. Mix!**

You can grow crystal-covered shapes and make them into a mobile. Or you can try to think of a material that will allow the crystals to grow on its surface. What would happen if you used a plastic toy? Or your brother's sock? Or your dog's collar? Try it!

And, hey, now that you know what a crystal looks like, be on the lookout for all the crystals in the world around you.

Crystalrama!